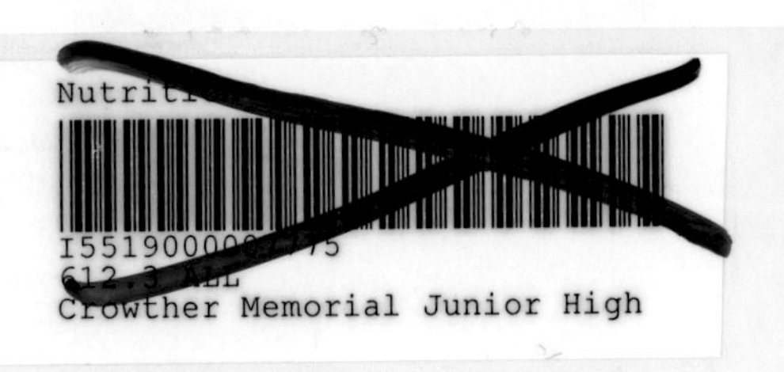

HEALTHY LIVING

Nutrition

ALEXANDRA POWE ALLRED

PERFECTION LEARNING®

Editorial Director: Susan C. Thies
Editor: Paula J. Reece
Design Director: Randy Messer
Book Design: Emily J. Greazel
Cover Design: Michael A. Aspengren

To my children—Kerri, Katie, and Tommy. You remind me every day why healthy living is so important. And we have fun, don't we?

Image credits:
© Associated Press: pp. 4, 9; © Tom Bean/CORBIS: p. 7, © Images.com/CORBIS: p. 25; © Sian Irvine/Stockfood America: p. 23; © North Wind Picture Archives: p. 8 (bottom)

Brand X Pictures: pp. 5, 26, 33; ClipArt.com: p. 8 (top); Corel Professional Photos: pp. 3, 28; Image100: p. 15; Perfection Learning Corporation: pp. 12, 13, 19 (bottom left), 20, 22, 30; Photos.com: back cover, front cover, pp. 1, 6, 16, 17, 19 (bottom right), 21, 24, 27, 29, 31, 32, 34; Rubberball: p. 11

A special thanks to the following for her scientific review of the book:
Dr. Marcie R. Wycoff-Horn, Assistant Professor, Department of Health Education and Health Promotion, University of Wisconsin-LaCrosse

Printed in the United States of America.
For information, contact
Perfection Learning® Corporation
1000 North Second Avenue, P.O. Box 500
Logan, Iowa 51546-0500.
Phone: 1-800-831-4190
Fax: 1-800-543-2745
perfectionlearning.com

1 2 3 4 5 6 PP 09 08 07 06 05 04

ISBN 0-7891-6425-6

Table of Contents

Introduction to Nutrition

Food is everywhere. Think about it. You can't go anywhere without seeing advertisements for food or hearing someone talk about a favorite food or drink. Fast-food restaurants appear around every corner. Magazine covers advertise the latest diet craze. Headlines in big, bold letters promise wonderful results: "Lose 15 Pounds in 30 Days!"

Food is very much a part of who you are and your history. But the value of food has changed. Your **ancestors** ate to survive. They grew and raised the food to feed their families.

The food they did not need to survive was traded for goods and services. Doctors would treat sick people for a pound of butter or a chicken. Items

like tea, sugar, and spices were exchanged for clothing, furniture, and animals.

Today, because food is so cheap and easy to get, Americans tend to think of it as something they *deserve* rather than something they *need*. Food is often offered as rewards. Have you ever gotten candy or an ice-cream cone because you brought home a good report card or cleaned your room?

With the change in attitude about food, Americans have forgotten how to eat. That may sound funny, but it's true. Americans supersize everything. They eat and drink too much.

People also make unhealthy food choices. They become overweight by living on junk food with little or no **nutritious** value. They may be making unhealthy choices because they don't know any better.

Just because people are thin does not mean they are healthy either. An adult who is 10 to 20 pounds over his or her ideal weight but who eats a healthy diet most likely is healthier than a thin person who does not eat healthy foods. A common mistake adults make is focusing on how they appear to other people rather than on how they feel. They become so worried about losing weight that, whether they are thin or heavy, they constantly diet and deny their bodies the proper **nutrients**. When you eat well-balanced, nutritious meals, you feel energized, strong, happy, and healthy. If you eat only junk food—as good as it might taste—you are slowly destroying your body.

Who knew food could be so complicated? The foods you choose to eat influence how well you do in school, how well you sleep, and how fast you run on the soccer field. So it's important to learn to eat healthy food as a young person. You will be happier and healthier, and you will be able to enjoy food so much more all of your life. *Bon appetite!*

1 How Americans Used to Eat

The early settlers in the colonies did not have nutritious diets. It wasn't because they were making unhealthy choices. It was because they could only eat what was in season and readily available. So their diet often lacked important **vitamins**.

A Good Reason to Drink Your Orange Juice!

A lack of vitamin C causes a disease called ***scurvy***. The symptoms of scurvy are frightening and painful. Teeth fall out, open sores appear on the legs and arms, and people become exhausted and sickly. Scurvy was common in the past but is rare in the United States today.

To keep food available all throughout the year was a challenge for the early settlers. Winters were particularly hard. They quickly had to learn how to **preserve** and store food.

The most important thing to keep from spoiling was meat. First the settlers **cured** it, usually by rubbing it with salt. This kept the meat fresh for a while, but not long enough to last the whole winter.

The meat had to be smoked if it was to be used later in the year. This was done in a smokehouse or even a fireplace or chimney. It usually took at least 40 hours to make the meat safe to store for later.

Early settlers built an outbuilding called a ***dairy*** to keep products such as milk and butter cool. Because they did not have refrigerators, the dairy was needed to keep food from spoiling. A dairy was built one or two feet below the ground, usually near a cool spring or creek. **Root cellars** were designed to keep vegetables such as potatoes, sweet potatoes, turnips, and carrots fresh. Like the dairy, the root cellar was dug into the ground, where the food could be kept cool.

Today, Americans have many modern conveniences, such as refrigerators, microwaves, and ovens. It is nearly impossible to imagine how early settlers once lived. There were no drive-through restaurants or four-minute microwave meals. Cooking meals for the family was difficult work.

Preparation for meals began as early as 5:00 in the morning. Children chopped firewood to build fires and carried heavy iron kettles filled with water. Ovens were nothing more than holes in the wall. There were no gauges to set the temperature. Young girls would stick their arm in the oven, above the flame, and try to count to 30. If they could not get to 30 before their skin started burning, they knew the oven was hot enough!

Root cellar

No Running to the Market for a Loaf of Bread!

The early settlers did not have the conveniences you have today. Even bread wasn't something to be taken for granted. They had to make all of their own food from scratch. No bread machines or box mixes for the pioneers! To make bread, they first had to make yeast, the substance that makes bread rise. Here is a recipe for yeast from 1883.

To make yeast:

Boil one pound of flour, a quarter of a pound of brown sugar, and a little salt, in two gallons water for one hour. When milk-warm, bottle it and cork it closed. It will be fit for use in 24 hours. One pint of it will make eighteen pounds of bread.

For early settlers, preparing food was an all-day chore. But once everyone was seated at the table, thanks and honor were given to those who made it all possible. No one ever ate out of boredom. There were very few overweight settlers. Eating was a privilege that came only after hard work and much time was given by everyone.

2

How Americans Eat Today

Today Americans have the means to safely and easily cook, freeze, and store food. But although they have better living conditions and more healthy foods than ever before, Americans are actually becoming less healthy. Why?

Do you feel like you have food on the brain? If it seems as if you think about food too much, perhaps it is because images of food are everywhere. Did you know the average American child watches more than 10,000 food commercials each year? But here is the unhealthy part. The commercials are not about eating fruits and vegetables. The commercials you are used to seeing show kids eating sugared cereals, fast food, soft drinks, and candy. Eating these foods and drinking these beverages seem really fun. People are dancing with cartoon characters. You can even win prizes if you eat or drink the right foods! Just seeing the commercials makes you want to be part of that fun, doesn't it?

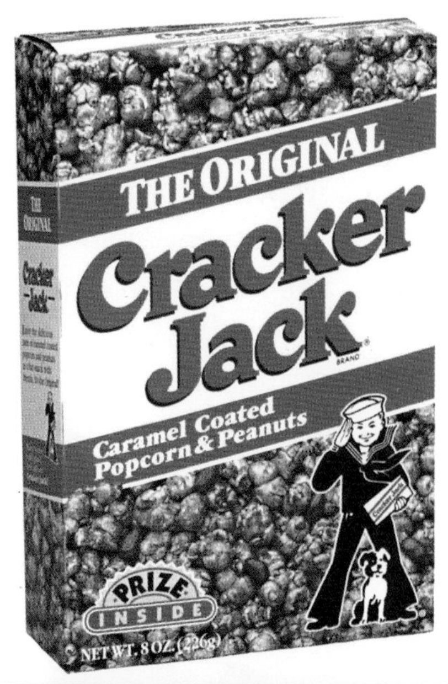

Together, doctors and scientists are discovering why Americans are becoming more and more overweight. These health experts report that American children have never been more overweight or out of shape in the history of this country. Since the 1960s, the number of overweight kids in the United States has nearly doubled. Health experts have ranked child and adult **obesity** as the number one health problem in this country. Did you know that over half the people living in the United States are overweight?

Obese vs. Overweight

Obese and overweight are not the same. One way to determine whether someone is considered obese or overweight is by using the body mass index, or BMI. To figure this, a doctor divides a person's weight by his or her height squared. The person's weight must be figured in kilograms and the height in meters. A child whose BMI is higher than the 97th percentile is considered obese. Kids are overweight if their BMI is between the 85th and 97th percentiles.

How It Breaks Down

Here's how a 13-year-old boy who is 5 feet, 3 inches tall would rank:
Above 141 pounds—obese
124–141 pounds—overweight
88–123 pounds—normal weight
Below 88 pounds—underweight

Obese children are likely to become obese adults. Children who are overweight can have **high blood pressure**, **asthma**, and **heart disease**. Obese children can also suffer from **type 2 diabetes**, a serious disease that can lead to blindness or **kidney failure**.

Simply put, when you consume more **calories** than you burn by working or exercising, you gain weight. As you have read, the early settlers worked from sunrise to sunset, burning many calories. But today, because people have remote controls, cars, and refrigerators, they don't have to (or want to) work nearly as hard as their ancestors did. Just since the 1970s, Americans have become less active. However, they consume 200 more calories a day. As a result, people are gaining weight.

McDonald's Says Good-Bye to Supersizes

In 2004, McDonald's stunned the fast-food industry by announcing that by the end of the year it would no longer include the Supersize option in its regular menu. The chain is also working on providing more fruit, vegetable, and yogurt options with its Happy Meals for kids.

Why Are Americans Eating More?

Americans eat too much, too fast. The United States is brimming with buffet restaurants where you get "all you can eat!" And how many times have you heard someone ask, "Would you like to make that bigger for just another 39 cents?" "Supersizing" has become common in fast-food restaurants.

Everyone likes a good deal. It feels like a bargain to get almost double the portions of food and/or drink for just a few cents more. But your body is getting more too. Too much more! In fact, a single supersized meal would likely give you more calories than you need for an entire day! If you are eating a supersized fast-food meal often, you can really add on more calories than you can possibly burn.

Supersized combo meals are not the only places where portion sizes have increased. A **portion size** is simply the amount of food a person chooses to eat. There are no standard sizes for portions. According to the World Health Organization, Americans eat double and sometimes triple the amount of food that people in other countries eat.

A **serving size**, on the other hand, is a standard amount that tells you how much to eat or shows you how many calories or nutrients are in a food. In the United States, people's ideas of what a serving size is have been changed by the **marketing** of larger food portions. For instance, in the 1950s, soft drinks were sold in small 6½-ounce bottles. Today Americans go to the convenience store and grab 20-ounce or even 34-ounce bottles. Research shows that in the 20 years between 1977 and 1996, hamburgers became 23 percent bigger. A plate of Mexican food in a restaurant got 27 percent bigger. And snacks, such as potato chips, pretzels, and crackers, grew by 60 percent!

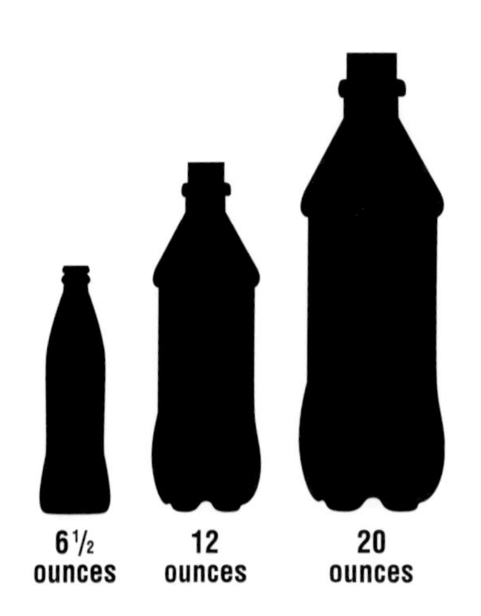

This chart shows how much bigger the average soft drinks are today than in the past.

Another reason Americans are eating more is because they see food advertised everywhere they go. Children are big targets for advertisers. Experts say that every year the food industry spends billions of dollars advertising to children. The problem is that most of the food advertised to kids isn't healthy. Half of all commercials aired during children's Saturday morning TV shows are for food. Kids spend more of their own money on food than anything else, including CDs, movies, clothes, or toys. They get so many messages about junk food that they can't really think about buying or eating anything else.

Try This!

For a quick and easy way to figure out serving sizes, just look at your hand!

Make a fist. This is about the size of 1 cup of food.

Now hold up your hand. Look at your palm. This is the average size of 3 ounces of meat, or one serving.

A handful is equal to about 1–2 ounces. Think about this when you grab a handful of chips, pretzels, or candy.

Now look at your thumb. It is about the size of 1 ounce of cheese. Two thumbs of cheese equals one serving.

The tip of your thumb is about the size of one teaspoon. Try to keep high-fat foods, such as peanut butter or mayonnaise, at one teaspoon.

1 cup

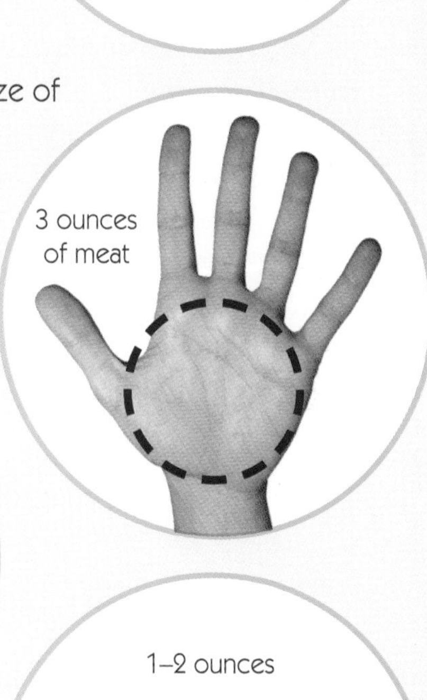

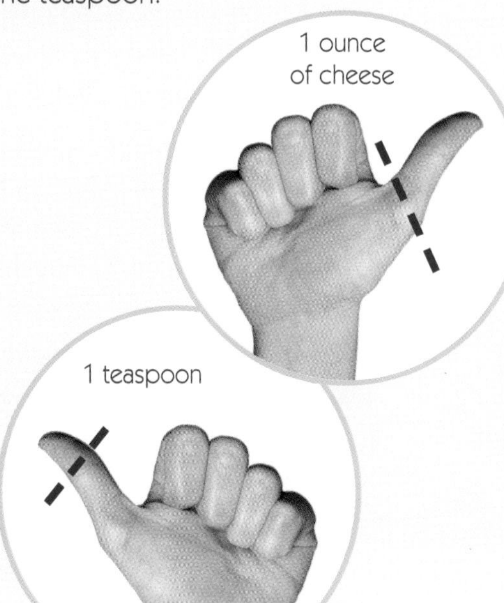

1–2 ounces

3 Understanding Fats, Carbohydrates, and Protein

Fat, **carbohydrates**, and **protein** give you energy, fight disease, build muscle, and rebuild body tissue. You need these nutrients to be healthy. You also need the vitamins and minerals you get from different foods. They help keep your body running like an efficient machine. These nutrients from food are absorbed by the **organs** in your body, fueling you for the day. You may have heard the expression "You are what you eat." This is definitely true!

The Skinny on Fat

You need to avoid fat because fat is bad, right? The truth is you need to have some fat in your diet. But most Americans digest too much fat. When cooked, fat has a good flavor and a smell most people like. Fat is what makes French fries crispy and helps make cookies chewy and gooey. That kind of fat, also known as **saturated fat**, can be unhealthy for you if you eat too much of it.

Have you ever watched someone drain the grease from a skillet into a cup or jar after frying bacon? That is fat. It takes on a solid form when kept at room temperature. **Unsaturated fats** are in liquid forms as olive oil, canola oil, and peanut oil, to name a few.

What does fat do for you? Fat is your largest fuel tank,

storing energy so you can play and work all day long. It helps you think better in school and at work, building brain cells and protecting your skin against weather that is too hot or too cold. Fat acts as a cushion for your organs. It protects your heart, kidneys, and everything else in your body from injury.

Eating too much fat, however, can make you overweight and lead to obesity and other health problems. You need fat—but a little bit goes a long way!

Those Complicated Carbs

Of all the things you feed yourself, carbohydrates (or "carbs") may be the most misunderstood. Carbohydrates come in two forms—simple and complex. Despite all you may have heard about limiting carbohydrates, they are extremely important for children and adults. They give you energy, help you think, and keep your body running smoothly. However, too many of the wrong kind of carbohydrates can leave you feeling empty and can turn to fat.

Simple carbohydrates, such as cakes and cookies, are mostly sugar. When you eat these foods, your body digests them very quickly. Your body will take these carbs and turn them into **glucose**. That means they will give you quick energy, but that energy won't last long.

The cells in your body that produce energy can only handle a certain amount of glucose at a time. When there is some left over, your cells store it for later. Some of it will fit in your muscles and liver. But whatever is left is turned into fat.

Processed foods with a lot of sugar, such as donuts or candy, will give your body more glucose than it can use. The extra is stored as fat. However, some simple carbs are good for you. Fruits, ice cream, and frozen yogurt will all give you quick energy in a healthy way without a lot of sugar.

Complex carbohydrates, however, are digested slowly and evenly. This leaves you feeling full and energized for a longer period of time. You can find these energy producers in bread, cereal, pasta, and rice. Vegetables such as cucumbers, corn, potatoes, tomatoes, carrots, lettuce, and peppers are also great places to get your complex carbs.

Powerful Protein

Protein builds muscles and fuels the development of organs in your body. It can help fight disease. Protein makes you strong. Unlike sugary foods that allow you to eat and eat, protein fills your stomach and keeps you satisfied longer. Also unlike sugar-based foods, excess protein in your body is not stored as fat. Instead, your body **excretes** it as waste. In other words, your body flushes the extra protein out when you go to the bathroom.

Protein is found in meat, chicken, fish, eggs, and nuts. It's also found in dairy products, like milk and cheese. Beans also contain protein.

4 Getting Acquainted with Vitamins and Minerals

Vitamins are organic substances, which means that they come from living things, such as fruits and vegetables. Vitamins prevent disease and help you grow. Some vitamins give you energy, and others help your blood clot. Unfortunately, your body can't make most vitamins. It relies on you to get what you need from the food you eat or the vitamin supplements you take.

Vitamins can be divided into two groups based on the way they are absorbed into your body. They are **fat-soluble** or **water-soluble**.

Vitamins A, D, E, and K are fat-soluble. They are dissolved into the body's fat or liver where they are stored. Some even stick around inside your body for up to six months! They come from meats, liver, egg yolks, oils, and some leafy green vegetables.

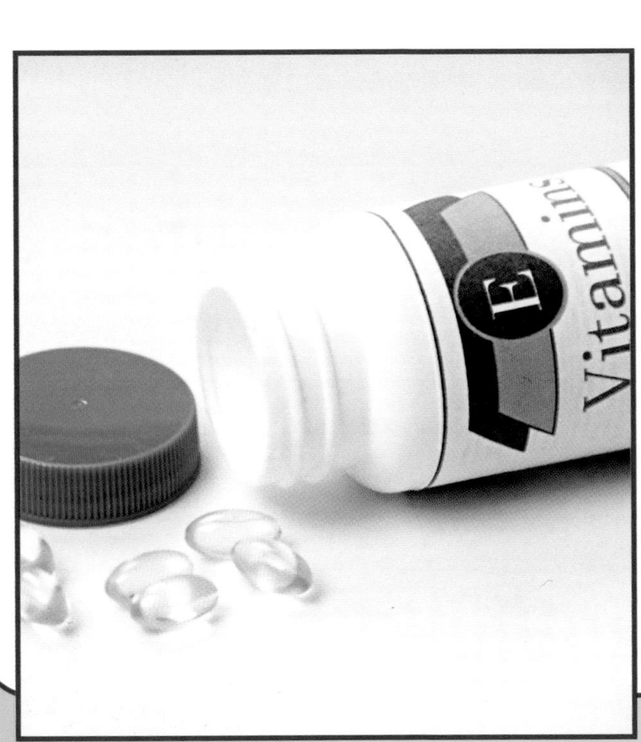

Water-soluble vitamins, such as C and B, come from watery foods like oranges, tangerines, and grapefruit. They are dissolved in water. This means that each time you go to the bathroom, the vitamins are flushed through your system. You must replace these vitamins often to stay healthy. These vitamins help your **immune system** fight diseases and infection.

Minerals come from nonliving things, such as water and soil. Yuck! Who eats dirt, right? Luckily, you can also get minerals from animals and plants as they absorb water, soil, and grass into their systems. The plants and animals pass these important nutrients on to you when you consume vegetables or meat.

Vitamin	What It Helps	Where You Can Find It
Vitamin A	Eyesight (even seeing in color), growth, healthy skin	Apricots, nectarines, carrots cantaloupe, spinach, pumpkins, fortified milk
B Vitamins (B1, B2, B6, B12, niacin, folic acid, biotin, pantothenic acid)	Make and use energy, make red blood cells (carry oxygen throughout the body)	Fish, meats (beef, pork, chicken), enriched breads and cereals, whole wheat grains, spinach, broccoli, dried beans
Vitamin C	Healthy gums, strong teeth, strong bones, strong muscles, healing of injuries, fighting off infection	Oranges, tangerines, lemons, grapefruit, honeydew melon, watermelon, strawberries, raspberries, orange juice, grapefruit juice, broccoli, tomatoes, green peppers
Vitamin D	Strong bones, strong teeth, absorption of calcium	Milk, pudding, tuna, eggs, the Sun
Vitamin E	Protects body's tissues, protects lungs, stores Vitamin A, blood clotting	Vegetable oils, corn, sunflower seeds, spinach, nuts
Vitamin K	Blood clotting	Broccoli, spinach, lettuce, cabbage, cheese

The Big and Small of It All

Macro means "large." Your body needs larger amounts of the macrominerals. *Micro* means "small," so—you guessed it! Your body needs these minerals in small amounts.

There are two different kinds of minerals in your body. **Macrominerals** are composed of **calcium**, phosphorus, potassium, sodium, chlorine, sulfur, and magnesium. You have probably heard about calcium because it helps to build stronger bones and fight **osteoporosis**. This is a condition that affects bones as you get older, causing them to be brittle and break easily. Calcium also promotes healthy teeth. The most common place to find calcium is in dairy products, such as milk or cheese. Some foods, such as orange juice, have calcium added to them to make them even more nutritious!

Microminerals are made up of minerals such as **iron**, fluoride, zinc, iodine, copper, manganese, chromium, selenium, boron, molybdenum, arsenic, nickel, and silicon. These are all needed in the human body as well. Iron is important because it is part of **hemoglobin**. This is the part of your red blood cells that carries oxygen from your lungs throughout your body. Every single part of your body needs oxygen to work properly. You can get your iron in red meat, baked beans, a baked potato with its skin, apricots, and even a common slice of bread!

What color is your food?

Did you know that the color of a fruit or vegetable can be important to your diet? **Phytonutrients**, or phytos, are what give fruits and vegetables their colors. The general rule is—the darker the color, the better. Blueberries, pink grapefruit (not white), sweet potatoes, and tomatoes produce **antioxidants**, which remove junk from your body. Dark green veggies are best in the vegetable department. Spinach greens are much healthier for you than the light green iceberg lettuce.

5

You Are What You Drink Too

What do you reach for when you are really thirsty? For many kids, the first thing they go for is a soft drink, or soda. In fact, 56 percent of eight-year-olds in America drink soda daily. And one-third of teenage boys drink at least three cans of soda each day! But what are you actually putting into your body when you chug a cola? Soft drinks are unhealthy both because of what they *contain* and what they *replace* in your diet.

The Problem with Pop

Soft drinks contain large amounts of sugar. One can or bottle of soda per day contains the daily recommended amount of sugar you need. And you haven't even eaten anything! When you think about everything else you eat in a day, you can see where your sugar intake may climb to unhealthy levels. Too much sugar can cause tooth decay, which can lead to cavities. Excess amounts of sugar can also cause you to gain too much weight.

Many soft drinks also contain a substance called ***caffeine***. It is actually a natural drug found in tea leaves, coffee beans, cocoa (used to make chocolate), and cola nuts (used to give some soft drinks their flavor). It stimulates the nervous system, making you feel more awake and alert. But it can also increase your heart rate.

Caffeine can also make you **dehydrated**. This means that your body gets weak because it doesn't have enough water. Drinks with caffeine in them make it harder for your body to keep enough water. So it's especially important on hot days or after you've exercised hard that you choose a drink without caffeine to quench your thirst.

An Ocean of Soft Drinks

In 2000, more than 15 billion gallons of soft drinks were sold in the United States. That figures out to be at least one can per day for every man, woman, and child!

Have you ever tried to talk to your mom, dad, or teacher before his or her first cup of morning coffee? This is not always a pleasant experience. That's because caffeine can also be **addictive**. That means that if you drink beverages with caffeine on a regular basis, your body will think it *needs* the caffeine for you to feel normal. Going without the caffeine can cause headaches, irritability, and just plain grumpiness. Depending on the level of addiction, after a few days or weeks of going "cold turkey" from caffeine, these withdrawal symptoms go away. Your body goes back to feeling normal without the drug's interference.

Soft Drinks vs. Milk

Soft drinks can be dangerous also because of what they *replace* in the diet, especially for kids. People who drink soft drinks are replacing nutritious foods and drinks with the soda. This means that they are missing out on key nutrients they need. One of these key nutrients is calcium. In the 1970s, boys drank more than twice as much milk as soft drinks, and girls drank 50 percent more milk than soft drinks. By 1996, both boys and girls drank twice as much soda as milk. Getting calcium is most important during the childhood and teen years. Bones are growing fast then, and calcium makes bones strong. Recent studies have shown that children who didn't drink enough milk had twice the number of bone fractures as children who drank milk. Not getting enough calcium can also lead to osteoporosis.

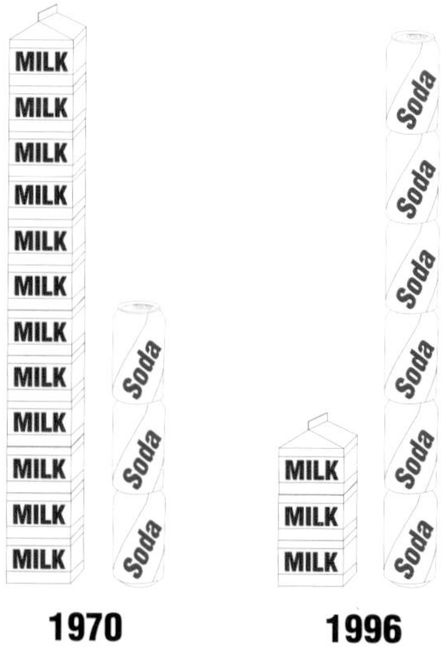

1970 1996

Got Strawberry Milk?

Some kids just don't like the taste of white milk. Does this sound like you? If so, you're in luck! Researchers have found that kids who drink flavored milk get more calcium than kids who choose other sugar-sweetened beverages, such as soft drinks. Doctors now recommend flavored milk as a healthy alternative to white milk. So grab an ice-cold glass of—*banana milk?*

Fruit Juice Frustration

But fruit juice has got to be good for you, right? Well, not necessarily. Fruit juice isn't *bad* for you. But it's not really that good for you either. The first thing you need to look at when you grab a bottle of juice is to see how much juice is actually in the drink. Many fruit juice drinks only contain 10 percent real fruit juice—or less! So what makes up the rest of the drink? Sugar is a main ingredient. And you have already learned how too much sugar is harmful to your body.

But what if you find a juice drink labeled "100% fruit juice"? It's got to be good for you, right? You're right. It is good for you—to a point. But drinking too much fruit juice can actually be bad for you. Fruit juice is filling, and if you drink enough of it, you won't eat other nutritious foods or drink nutritious drinks such as milk. Fruit juices have some vitamins and minerals, but they really don't have as much as the fruits themselves. Your body will benefit much more if you eat an apple rather than drink a glass of apple juice. So go ahead and have a glass of 100% orange juice. But don't overdo it!

Fruit Juice 101

- Orange juice has the highest amount of vitamin C and potassium and is a good source of other vitamins. It also contains cancer-fighting substances.
- Grapefruit juice has the second highest amount of vitamin C.
- Apricot nectar is high in vitamin A and has a small amount of iron and zinc.
- Prune juice is highest in iron, zinc, and **fiber**.
- White grape juice is high in vitamin C and is the best juice for healing the intestines.
- Freshly squeezed juice contains more vitamin C than canned or frozen juices that are made from concentrate.
- A general rule is—the cloudier the juice, the more nutritious it is. Juices that you can see through contain mostly water.

Water, Water Everywhere

So what about water? Is it really that good for you? Definitely! In fact, you need water to survive. But what's so great about it?

First of all, water is the main ingredient in all of the fluids in your body, including your blood. These fluids are responsible for traveling throughout your body, helping you digest food and carrying both nutrients and waste. Water also keeps your joints lubricated so your bones don't rub together. This helps your body move smoothly and keeps you on the go!

Water is also responsible for keeping you cool. Do you like to sweat? Most people don't. But sweat is actually necessary for cooling your body. When your body gets hot, water comes up through your skin as sweat. The sweat then evaporates, or disappears, into the air. When it evaporates, it cools your skin. This causes your blood and your insides to cool down too. If your body doesn't have enough water in it to produce sweat, you will overheat. This can lead to dehydration or heat exhaustion.

So, you need water every day to make your body work like it should. On average, you lose 2 to 3 quarts of water a day sweating and going to the bathroom. If you're really hot or exercising a lot, you'll lose even more. It's essential that you replace this water. You'll get some water from the foods you eat, especially fruits and vegetables. But you should also drink water too—6 to 8 cups a day. Don't wait until you're thirsty, either. This is your body's way of telling you it's getting low on water. Make it a habit to drink a glass of water on a regular basis. The next time you have a choice between drinking something sweet or watering yourself, take a big swig of water!

6 The Food Guide Pyramid

Now that you know what kinds of food are good for you, how do you know *how much* of each you should eat every day? An easy way to tell is by looking at the Food Guide Pyramid. It shows you the groups of foods that you need to eat for a healthy diet. Because it is shaped like a pyramid, with the bottom being bigger than the top, it is easy to tell which foods you should make a bigger part of your menu.

Bread, Cereal, Rice, and Pasta Group

Since this group is on the bottom of the pyramid, that means these foods should make up the biggest part of your daily diet. You'll get most of your energy from this group because these foods are full of carbohydrates. The pyramid suggests that you eat 6 to 11 servings per day. This sounds like a lot. Take a look at what makes up one serving. You'll see that you really won't have to eat a whole loaf of bread to meet the requirement! In fact, eat a whole English muffin and a bowl of cereal at breakfast, and you'll get 3 servings. Eat a whole sandwich at lunch, and you'll get 2 more. You're almost there—and it's only noon!

What's in a serving?

- 1 slice of bread
- ½ cup of cooked rice or pasta
- ½ ounce of cooked cereal
- 1 ounce of cold cereal
- ½ bagel or English muffin

Vegetable Group

By now you should know why some grown-up is always telling you to "Eat your veggies!" Vegetables are loaded with all different kinds of vitamins and minerals. Vegetables also contain complex carbohydrates, which give your body energy. Don't forget that the darker the color of the vegetable, the better it is for you. The food pyramid says you should eat 3 to 5 servings of vegetables a day.

What's in a serving?

- 1 cup of raw leafy vegetables
- ½ cup of other raw vegetables or cooked vegetables
- ¾ cup of vegetable juice

Explorer and—Nutrition Expert?

In 1772, something happened that changed people's diets and saved many lives. Captain James Cook, an English sailor and explorer, began a three-year voyage around the world. Before he set off, he made sure to have a good supply of fresh fruits onboard his ships. Each time they stopped at a new port, Cook restocked the supply. Guess what happened? Not one of Cook's sailors became ill in the three years they traveled. This was such great news that from that time on, many ships were well stocked with fruits and vegetables. And people who stayed on land saw the importance of fresh fruits and vegetables for their health too.

Fruit Group

Like vegetables, fruits are full of vitamins and carbohydrates. They are also excellent sources of fiber. Fiber is important in helping your body digest food, and it helps protect you from getting sick. According to the Food Guide Pyramid, you should eat 2 to 4 servings of fruit every day. Eat one serving at breakfast, lunch, and dinner, along with one as a snack, and you're set!

What's in a serving?

- 1 medium-sized apple, banana, or orange
- 1/2 cup of cooked or canned fruit
- 1/2 cup of fruit juice

Milk, Yogurt, and Cheese Group

Take a look at where this group is on the pyramid. It's close to the top. This means that even though these foods are important, you don't need as much of them as you do the foods closer to the bottom of the pyramid. But this group is especially critical because it is the main source of calcium. It will also give you protein, along with vitamins you need. You should get 2 to 3 servings of this group per day.

What's in a serving?

- 1 cup of milk
- 1 cup of yogurt
- 1 ounce of cheese

Meat, Poultry, Fish, Dry Beans, Eggs, and Nuts Group

All of the foods in this group give you a very important nutrient—protein. They also provide you with other minerals, such as iron and zinc. The pyramid suggests you eat 2 to 3 servings of this important group each day. This is easy to do when you consider one serving of meat is only the size of a deck of cards!

What's in a serving?

- 2 to 3 ounces of cooked lean meat, poultry, or fish
- 1 egg
- ½ cup of cooked dry beans (pinto, black, or navy)
- 1 tablespoon of peanut butter

Fats, Oils, and Sweets

Since these are perched at the top of the pyramid, your body should have less of these than foods from the other groups. Your body needs some fat, but it's pretty easy to get too much fat in one day. Sweets give you energy because they contain simple carbohydrates, but this energy is short-lived. Sweets also contain a lot of sugar and calories, which aren't good for you. You don't have to cut out these foods altogether. Just eat a little and not all the time. Make sure you balance with all of the other foods in the groups that will make you feel and look your very best!

7 Developing Healthy Habits

Now that you know what to eat and how much of it, get in the habit of keeping your health in mind when you dine! There are some things you can do to start developing healthy eating habits.

Give Yourself a Good Start

Have you ever heard anyone say, "Breakfast is the most important meal of the day"? Well, this time you can believe what you hear. Think about what time you usually eat your last meal or snack in the evening. Now think about what time you get up in the morning. Most likely it's been 10 hours or so since you've eaten. That's a long time! Your brain and muscles need some energy to get thinking and moving. To get this energy, they need *you* to eat some healthy food. Imagine how long your body goes without any food when you skip breakfast and don't eat until lunch. No wonder you feel sleepy and find it hard to think on those days!

It's important to always eat breakfast.

The best boost you can give your body in the morning is to eat a variety of healthy breakfast foods. A couple of donuts and a can of soda aren't going to do the trick! Take a look at the food pyramid and try to fit as many servings of different groups as you can. That way, your body will get a jump-start with a variety of vitamins, minerals, and nutrients.

Eat Smart at School

Do you pack your own lunch or eat what the school serves each day? Maybe you mix it up depending on the menu. Even if you eat at school every day, don't assume that you're getting only the most nutritious food. There are regulations about what schools can serve in the cafeteria. However, recent studies show the typical school lunch is still higher in fat than it should be. And most schools offer choices. You may be able to buy a slice of pizza, French fries, a candy bar, and a can of soda every day for lunch if you want. As you now know, this lunch is not a healthy choice.

If you choose to bring your own lunch, think about your health when you pack your bag. Try to get a balanced lunch, including as many of the groups on the food pyramid as you can. Instead of a candy bar, throw in an apple or a peach. For a sandwich, buy whole-grain bread instead of white bread. And resist the urge to pack a bottled soft drink. Instead, fill a thermos with milk or just drink water. Eating a healthy lunch will make you less tired and better able to concentrate for the second half of your school day!

What May Be Lurking in the Cafeteria

According to the U.S. Department of Agriculture, 78 percent of schools offer soft drinks in their cafeteria. Salty snacks and high-fat treats are offered in 64 percent of schools.

Slow Down

Rush, rush, rush. Everyone is in a rush. This is especially true at mealtimes today. But you should always take time for meals. Slow down and sit down while you eat. Do not stand at the kitchen counter. Studies have shown that people tend to eat much faster and much more when standing. Let your digestive system do its work while you sit and enjoy a meal slowly. Here are some fun ways to slow down at the dinner table.

- Try using chopsticks to handle your food.
- Let 1 minute pass before each new bite.
- Hold your fork in the opposite hand than the one with which you usually eat.
- Put down your fork each time you have taken a bite and do not pick it up again until you have chewed and swallowed the bite.

Each meal should last at least 20 minutes. Sometimes, like at school, you may not have a choice of how much time you have. But you can at least try to slow down at dinner. If you are with friends or family, the meal should last 40 minutes. This way you know you are eating slowly, talking and listening, and truly enjoying the food. Try cutting your food into little pieces. Remember, the bigger the bites, the harder it is for your body to break up and digest the food.

Look at the Clock

Are you really hungry? Or are you just eating because of something you saw on television? Are you bored? Restless? Listen to your body and determine if you really are hungry. Look at the clock to see what time it is. If it is late in the afternoon, make yourself a light snack so you won't spoil your dinner. Or just have a glass of water. It will satisfy your urge to eat or drink something, and it will be good for you.

Bring the Family Back to the Table

Your older sister has soccer practice on Mondays and Thursdays, your little sister has violin practice on Tuesdays, and you have basketball on Tuesdays and Fridays. Your parents don't get home until 6:00 every evening, and everyone is running around like crazy. Trying to get everyone at the table at the same time is nearly impossible. Nearly—but not completely.

In the beginning, there may only be time for one or two meals together each week. Sit down with your family, work out a plan, and stick to it. Maybe you have decided to eat together every Wednesday and Saturday or every Thursday and Sunday. Then plan together as a family and divide the chores. Perhaps you can make the salad while your sister sets the table and your mother handles the hot food. However your family creates your meal, have some set rules before everyone takes a seat.

10 Rules for the Table

1. Only healthy drinks will be served. No sugary drinks allowed!
2. Limit the amount of salt used. Experiment with other seasonings for meats and vegetables.
3. Think like a rabbit! Try to begin your meals with a small salad. By eating a small salad before the main meal is served, you will get much-needed minerals and vitamins. You will be less likely to fill up on the heavier foods.

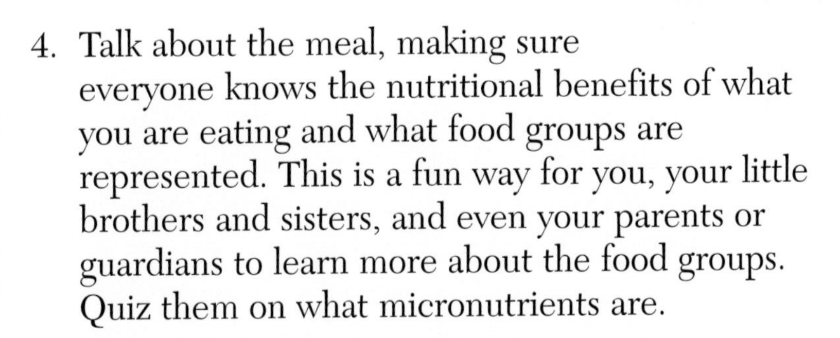

4. Talk about the meal, making sure everyone knows the nutritional benefits of what you are eating and what food groups are represented. This is a fun way for you, your little brothers and sisters, and even your parents or guardians to learn more about the food groups. Quiz them on what micronutrients are.
5. No negative conversations at the dinner table. This means no tattling, no arguing, and no complaining!
6. Set a timer for 30 to 40 minutes. This is the time your family should have together at the table. Even if you need to bring notes to the table for things to talk about, be sure the family sits together until the timer bell rings. Soon enough, your family will want to sit together longer than the timer says.

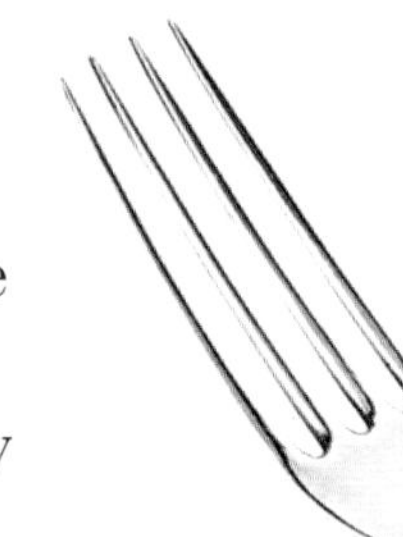

7. Practice eating slowly together. For fun, have your family members use their forks with the opposite hands than they usually use. Make them count how many times they chew.

8. Take turns talking about how your day was. When it is someone else's turn to talk, use this time to be a good listener and ask questions. You will be amazed what you learn about your family during this time.

9. Discuss what the next meal should be, letting everyone have a vote. Be careful! Don't let your little sister ruin your good efforts by ordering a pizza. Be sure the adults control the vote by ensuring that vegetables, protein, and carbohydrates are presented.

10. Talk about one thing outside your family. This could be a news story about another person or place. Perhaps you and your family members could take turns telling about something in the newspaper or on TV or that you learned at school or work. Older family members can help younger ones so everyone has a turn. Good food and good conversation go together.

America is a country obsessed with food. Getting to know your body and how the right foods make it run better will turn this obsession with food from a negative trait into a good thing!

Internet Connections and Related Readings for *Nutrition*

KidsHealth (www.kidshealth.org/kid)
This is the one-stop shop for the answers to all of your questions about your health as it relates to nutrition. Read articles written just for you, such as "The Food Guide Pyramid." Play the Mission Nutrition game. Visit the glossary to find quick definitions of all sorts of medical terms you need to know. The "Kids' Health Problems" page offers a multitude of information on health problems you or someone you know may have, from allergies to infections to cancer. "The Game Closet" offers a health quiz, a body scavenger hunt, rainy day fun, and road trip fun. This Web site gives you straight talk about health meant just for kids.

Nutrition Explorations: Kids
(www.nutritionexplorations.org/kids)
Read all about nutrition, such as "How Much Should You Eat?" Tour the Food Guide Pyramid to learn all about healthy servings. Be entertained with activities, such as downloading nutrition stickers or playing Monster Nutrition, an interactive game. Become a Breakfast Detective, reading the clues and meeting the food challenge. Make a virtual milk shake and answer Food Riddles. To monitor how healthy your food choices are, fill out the Nutrition Tracker, where you can record your meals and compare your servings.

Food and Health by Enid Fisher. Good Health Guide series. Gareth Stevens, 1998. [RL 4.9, IL 3–7] (5882606 HB)

Food Pyramid by Joan Kalbacken. Introduces the food pyramid describing each level in detail and discussing the benefits of healthy eating. True Books—Food & Nutrition. Children's Press, 1998. [RL 4, IL K–4] (3775106 HB)

Food Rules! The Stuff You Munch, Its Crunch, Its Punch, and Why You Sometimes Lose Your Lunch by Bill Haduch. A comprehensive book on food and nutrition created for young readers. Dutton, 2001. [RL 5.8, IL 3–8] (3765606 HB)

Vitamins and Minerals by Joan Kalbacken. Introduces the major vitamins and minerals found in various foods and discusses them in relation to nutrition. Children's Press, 1998. [RL 4, IL K–4] (3777506 HB)

•RL = Reading Level
•IL = Interest Level
Perfection Learning's catalog numbers are included for your ordering convenience. *HB* indicates hardback.

Glossary

addictive (uh DIK tiv) causing a person to be dependent upon a substance

ancestors (AN ses terz) people in a family from whom a person is descended, or came from

antioxidants (an tee AWKS uh duhnts) substances thought to be effective in helping to prevent disease

asthma (AZ muh) disease that causes tightening of the chest, coughing, and difficulty in breathing

caffeine (kaf FEEN) substance found naturally in coffee, tea, cola nuts, and cocoa that stimulates the central nervous system

calcium (KAL see uhm) **macromineral** that helps build strong bones (see separate entry)

calories (KAL uh reez) the measure of how much energy food supplies to the body

carbohydrates (kar boh HY drayts) sugars and starches the body makes to produce energy

complex carbohydrates (kom PLEKS kar boh HY drayts) starchy **carbohydrates** that are digested slowly and evenly for more sustained energy (see separate entry)

cured (kyurd) prepared by processing so it will last longer

dairy (DAIR ee) a room or building where milk and milk products are kept cool

dehydrated (dee HY drayt ed) having lost too much water or body fluids so that bodily organs begin to shut down

excretes (eks KREETS) eliminates waste from the body

fat (fat) the body's major form of energy storage

fat-soluble (fat SOL yuh buhl) dissolved in the body's **fat** or liver (see separate entry)

fiber (FEYE ber) food material in human food that can't be digested and so stimulates the intestines

glucose (GLOO kohs) sugar substance that is carried in the blood and gives a body energy

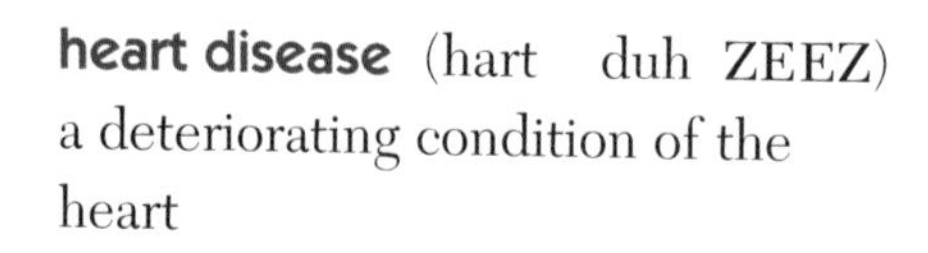

heart disease (hart duh ZEEZ) a deteriorating condition of the heart

hemoglobin (HEE muh gloh buhn) a substance in red blood cells that transports oxygen

high blood pressure (heye bluhd PRES sher) too much pressure in the arteries that can lead to heart problems

immune system (im MYOON SIS tuhm) body system that creates a defense against foreign substances

iron (EYE uhrn) a **micromineral** essential to the body that makes up **hemoglobin** (see separate entries)

kidney failure (KID nee FAYL yer) the shutting down of the organs that filter liquid waste from the blood and help regulate the amount of water and other substances in the body

macrominerals (mak roh MIN uh ruhlz) minerals such as **calcium**, phosphorous, and magnesium, of which the body needs large amounts (see separate entry)

marketing (MAR kuht ing) using advertising to attract buyers to a product for sale

microminerals (meye kroh MIN uh rulhz) minerals such as **iron**, fluoride, zinc, iodine, copper, manganese, chromium, selenium, molybdenum, arsenic, boron, nickel, and silicon that are needed in the body in smaller amounts (see separate entry)

nutrients (NOO tree uhnts) substances that provide nourishment to the body

nutritious (noo TRISH uhs) having good nutritional value; healthful

obesity (oh BEE suh tee) a condition of having excessive body fat

organs (OR guhnz) bodily parts that perform functions

osteoporosis (os tee oh puh ROH suhs) a condition that causes bones to become brittle and break easily

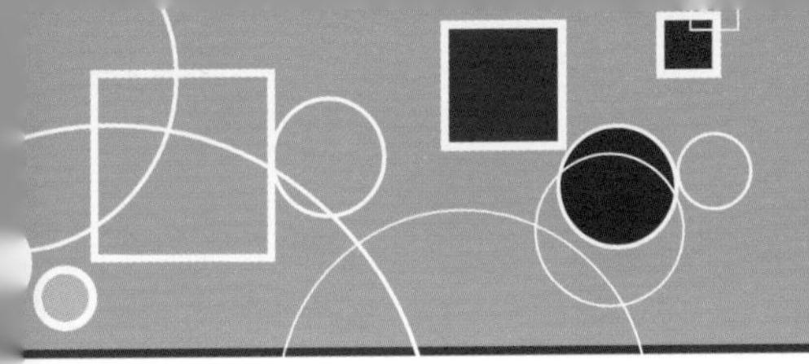

phytonutrients (feye toh NOO tree uhnts) naturally occurring substances found in plants that contribute to their color

portion size (POR shuhn seyez) the amount of food a person chooses to eat

preserve (pri ZERV) to can, pickle, or similarly prepare food for future use

protein (PROH teen) substance a body needs to build muscles, fuel development of the **organs** and fight off disease (see separate entry)

root cellars (root SEL luhrz) pits used for storing fruits and vegetables

saturated fat (SACH uh ray ted fat) **fat** that comes mainly from animal foods and can be harmful if a person eats too much, too often (see separate entry)

scurvy (SKER vee) a disease that caused teeth to fall out and bleeding due to a lack of vitamin C

serving size (SER ving seyez) a standard amount that tells a person how much to eat or how many calories or **nutrients** are in a food (see separate entry)

simple carbohydrates (SIM puhl kar boh HY drayts) **carbohydrates** that are digested quickly for short-lived energy and are more likely to be stored as **fat** (see separate entries)

type 2 diabetes (teyep 2 deye uh BEE teez) serious disease that is the result of the body not making enough insulin, a hormone that helps get glucose, or sugar, into the cells of the body to be used for energy

unsaturated fat (UN sach uh ray ted fat) **fat** that comes mainly from plants and fish that is necessary for the body (see separate entry)

vitamins (VEYE tuh muhnz) organic substances that, when absorbed into the body, prevent disease and help promote growth

water-soluble (wah ter SOL yuh buhl) dissolved in water and quickly eliminated

Index